CHEMICAL INTERACTIONS
LAB NOTEBOOK

DEVELOPED AT LAWRENCE HALL OF SCIENCE, UNIVERSITY OF CALIFORNIA AT BERKELEY
PUBLISHED AND DISTRIBUTED BY DELTA EDUCATION

FOSS Middle School Staff and Associates

FOSS Middle School Curriculum Development Team

Linda De Lucchi and Larry Malone, Co-Directors
Teri Dannenberg, Curriculum Developer; **Ann Moriarty**, Curriculum Developer
Dr. Susan Ketchner, Content Specialist; **Susan Kaschner Jagoda**, Curriculum Developer
Virginia Reid, Curriculum Consultant; **Dr. Terry Shaw**, Curriculum Consultant
Dr. Kathy Long, Assessment Coordinator
Carol Sevilla, Publications Coordinator; **Rose Craig**, Illustrator
Alev Burton, Program Assistant

ScienceVIEW Multimedia Design Team

Alana Chan, Project Manager; **Leigh Anne McConnaughey**, Principal Illustrator
Dan Bluestein, Lead Programmer; **Chris Cianciarulo**, Programmer
Roger Vang, Programmer; **Nicole Medina**, Production Assistant

Special Contributors

Marshall Montgomery, Materials Design; **John Quick**, Photography and Video
Barbara Stebbins, 7th-grade teacher, Black Pine Circle School, Berkeley, CA, and her students
Advisers: **Jennifer Claesgens**, **John Sheridan**, **Dr. Angelica Stacy**, **Lillie Tucker Akin**
Content Reviewer: **Dr. Dorothy Gabel**

Delta Education FOSS Middle School Team

Bonnie Piotrowski, FOSS Editorial Director; **Mathew Bacon**, Vice President, Product Development
Project Team: **Megan Dumm**, **Tom Guetling**, **Joann Hoy**, **Haley January**, **Lisa Lachance**,
Angela Miccinello, **Cathrine Monson**, **John Prescott**, **Barbara Resch**, **Jennifer Roehrig**, **Nina Whitney**

National Trial Teachers

Seth Corrigan, M.L. King Middle School, Berkeley, CA
Michael Romero, Angerine Middle School, Boulder, CO
Stephen "Doc" Holaday, Abilene MS, Abilene, KS
Jessica Penchos, Timilty MS, Roxbury, MA
Cheryl Hall, Grover Cleveland MS, Dorchester, MA
Sarah Chapin and Joanna Snyder, Hudson High School, Hudson, MA
Gayle Dunlap, Saddle Brook Middle School, Saddle Brook, NJ
Corinne De Keukela, Walter T. Bergen MS, Bloomingdale, NJ
Andrew Kramar, Cavallini Middle School, Upper Saddle River, NJ
Beata Ferris, Belle Fourche School District, Belle Fourche, SD
Steven Barry, Spring Branch MS, Houston, TX
Charlene D. Guerra, Landrum MS, Houston, TX
Diane Mayer and Jason A. Schmid, Oak Harbor MS, Oak Harbor, WA
Jack Aldridge, Thurgood Marshall MS, Olympia, WA
Melissa Barnard and Anna Horton, Nelsen MS, Renton, WA
Lori Aegerter and Sheryl Dhuey, Cherokee Heights MS, Madison, WI
Suzanne M. Folberg and Mary E. Joe, O'Keeffe MS, Madison, WI

LHS

FOSS for Middle School Project
Lawrence Hall of Science, University of California
Berkeley, CA 94720 510-642-8941

Delta Education
...because children learn by doing.®

Delta Education, LLC
P.O. Box 3000 80 Northwest Blvd.
Nashua, NH 03063
1-800-258-1302

The FOSS Middle School Program was developed in part with the support of the National Science Foundation Grant ESI 9553600. However, any opinions, findings, conclusions, statements, and recommendations expressed herein are those of the authors and do not necessarily reflect the views of the NSF.

121-

Chemical Interactions
1 2 3 4 5 6 7 8 9 BAB 09 08 07 06

ISBN-10: 1-58356-4
ISBN-13: 978-1-58356-4

Chemical Interactions Lab Notebook

Table of Contents

Investigation 6: Heat of Fusion

Investigation 7: Phase Change

Investigation 8: Solutions

Investigation 9: Reaction

Investigation 10: More Reactions

CHEMISTRY GLOSSARY

WARNING — This set contains chemicals that may be harmful if misused. Read cautions on individual containers carefully. Not to be used by children except under adult supervision.

Name ______________________________

Period ____________ Date ________________

MYSTERY MIXTURE

Materials

- • 5-mL sample of mystery mixture
- 1 Plastic cup, 250-mL
- 2 Pipettes
- 2 Hand lenses
- 1 Cup of water
- • Protective eyewear

Part 1. Observe the unknown mixture.

a. Put on your protective eyewear.

b. Put one 5-mL spoon of the mystery mixture into a cup.

c. Observe the mixture. (Do not touch or taste the mixture.)

d. Record your observations.

____

____

____

____

____

Part 2. Add water.

a. Add one pipette of water to the mystery mixture in the cup. Do not use the pipette to stir the mixture.

b. Observe. Take turns putting additional pipettes of water into the cup. Observe.

c. Record your observations.

____

____

____

____

____

WARNING — This set contains chemicals that may be harmful if misused. Read cautions on individual containers carefully. Not to be used by children except under adult supervision.

Name ______________________________

Period ____________ Date ____________________

WHITE SUBSTANCE INFORMATION

Fill in the chart with the information requested for each white substance.

Chemical name	Chemical formula	Common name	Observations	Uses

Look for patterns in the chemical names and chemical formulas for the substances. What do you see?

____

____

____

____

WARNING — This set contains chemicals that may be harmful if misused. Read cautions on individual containers carefully. Not to be used by children except under adult supervision.

Name ______________________________

Period ____________ Date ______________

MYSTERY-MIXTURE ANALYSIS

Materials

- 1 Set of nine white substances
- 1 Vial of mystery mixture
- 2 Dropper bottles of water
- 10 Minispoons, green
- 2 Well trays
- • Protective eyewear

Challenge

Find out which two substances are in the mystery mixture.

Procedure

a. Put one level minispoon of two different substances (or two minispoons of one substance) in a well. Note the number of the well.

b. Add 10 drops of water. Observe and record.

Well number	Substance 1	Substance 2	Results
1	Mystery mixture		
2	+		
3	+		
4	+		
5	+		
6	+		
7	+		
8	+		
9	+		
10	+		
11	+		
12	+		
	+		
	+		

WARNING — This set contains chemicals that may be harmful if misused. Read cautions on individual containers carefully. Not to be used by children except under adult supervision.

Name ______________________

Period ____________ Date ______________

MYSTERY-MIXTURE SUMMARY

Well	Substances	Description of fizzing	Other observations	Large-scale reactions
1	Ascorbic acid + calcium carbonate $C_6H_8O_6 + CaCO_3$			
2	Ascorbic acid + sodium bicarbonate $C_6H_8O_6 + NaHCO_3$			
3	Ascorbic acid + sodium carbonate $C_6H_8O_6 + Na_2CO_3$			
4	Calcium chloride + sodium bicarbonate $CaCl_2 + NaHCO_3$			
5	Citric acid + calcium carbonate $C_6H_8O_7 + CaCO_3$			
6	Citric acid + sodium bicarbonate $C_6H_8O_7 + NaHCO_3$			
7	Citric acid + sodium carbonate $C_6H_8O_7 + Na_2CO_3$			
8	Mystery mixture			

Identify the two substances in the mystery mixture and explain how you identified them.

____

____

____

Name ______________________________

Period ____________ Date ________________

MYSTERY-MIXTURE ELEMENTS

Part 1. Identifying elements

Substance	Chemical formula	Elements
Calcium carbonate	$CaCO_3$	
Sodium carbonate	Na_2CO_3	
Sodium bicarbonate	$NaHCO_3$	
Magnesium sulfate	$MgSO_4$	
Calcium chloride	$CaCl_2$	
Sodium chloride	$NaCl$	
Ascorbic acid	$C_6H_8O_6$	
Citric acid	$C_6H_8O_7$	
Sucrose	$C_{12}H_{22}O_{11}$	

Part 2. Questions

1. Which substance has the greatest number of elements? ____________ How many? _____
2. Altogether, how many different elements are in the nine substances? _________
3. Which element is found in the greatest number of substances? ______________
4. How many elements are in the substance carbon dioxide? _________________
5. How many elements are in the substance water? _________________________
6. Which of the nine substances are made of two elements? __________________________

   ____
7. Which of the nine substances are made of three elements? ________________________

   ____
8. Which of the nine substances are made of four elements? _________________________

   ____

Name ______________________________

Period ____________ Date ________________

ELEMENTS QUESTIONS

SRB page 8

1. What is an element?

2. How are matter and elements related?

3. How was Mendeleyev able to predict the existence of elements that had not yet been discovered?

4. What is the periodic table of the elements?

Name ____________________

Period ____________ Date ____________

ELEMENTS IN PRODUCTS

Part 1. List the elements found in several products.

Product name	Elements

Part 2. Analyze the elements in the products.

1. How many different elements did you find in all the products you investigated?

____

2. What is the most common element in the products you investigated?

____

3. How many metals did you find in the products you investigated? List them.

____

____

Name ______________________________

Period ____________ Date ________________

PERIODIC TABLE

The Periodic Table of the Elements

1 H Hydrogen																	2 He Helium
3 Li Lithium	4 Be Beryllium											5 B Boron	6 C Carbon	7 N Nitrogen	8 O Oxygen	9 F Fluorine	10 Ne Neon
11 Na Sodium	12 Mg Magnesium											13 Al Aluminum	14 Si Silicon	15 P Phosphorus	16 S Sulfur	17 Cl Chlorine	18 Ar Argon
19 K Potassium	20 Ca Calcium	21 Sc Scandium	22 Ti Titanium	23 V Vanadium	24 Cr Chromium	25 Mn Manganese	26 Fe Iron	27 Co Cobalt	28 Ni Nickel	29 Cu Copper	30 Zn Zinc	31 Ga Gallium	32 Ge Germanium	33 As Arsenic	34 Se Selenium	35 Br Bromine	36 Kr Krypton
37 Rb Rubidium	38 Sr Strontium	39 Y Yttrium	40 Zr Zirconium	41 Nb Niobium	42 Mo Molybdenum	43 Tc Technetium	44 Ru Ruthenium	45 Rh Rhodium	46 Pd Palladium	47 Ag Silver	48 Cd Cadmium	49 In Indium	50 Sn Tin	51 Sb Antimony	52 Te Tellurium	53 I Iodine	54 Xe Xenon
55 Cs Cesium	56 Ba Barium	71 Lu Lutetium	72 Hf Hafnium	73 Ta Tantalum	74 W Tungsten	75 Re Rhenium	76 Os Osmium	77 Ir Iridium	78 Pt Platinum	79 Au Gold	80 Hg Mercury	81 Tl Thallium	82 Pb Lead	83 Bi Bismuth	84 Po Polonium	85 At Astatine	86 Rn Radon
87 Fr Francium	88 Ra Radium	103 Lr Lawrencium	104 Rf Rutherfordium	105 Db Dubnium	106 Sg Seaborgium	107 Bh Bohrium	108 Hs Hassium	109 Mt Meitnerium	110 Ds Darmstadtium	111 Rg Roentgenium	112 Uub	113 Uut	114 Uuq	115 Uup	116 Uuh	117 Uus	118 Uuo

57 La Lanthanum	58 Ce Cerium	59 Pr Praseodymium	60 Nd Neodymium	61 Pm Promethium	62 Sm Samarium	63 Eu Europium	64 Gd Gadolinium	65 Tb Terbium	66 Dy Dysprosium	67 Ho Holmium	68 Er Erbium	69 Tm Thulium	70 Yb Ytterbium
89 Ac Actinium	90 Th Thorium	91 Pa Protactinium	92 U Uranium	93 Np Neptunium	94 Pu Plutonium	95 Am Americium	96 Cm Curium	97 Bk Berkelium	98 Cf Californium	99 Es Einsteinium	100 Fm Fermium	101 Md Mendelevium	102 No Nobelium

Name ______________________________

Period ____________ Date ________________

RESPONSE SHEET—ELEMENTS

Carla studied the ingredients on a box of cereal. She made a list of the elements she found. She told her friend,

This cereal contains eight different elements. I wonder what the rest of the cereal is made of.

If you were Carla's friend, what would you tell her?

____

____

____

____

____

____

____

____

Name ______________________________

Period __________ Date ______________

ELEMENTS IN THE UNIVERSE QUESTIONS

SRB page 13

1. What element is among the five most abundant elements in the Sun, Earth, ocean, atmosphere, and organisms?

2. What does it mean when people say everything is made of stardust?

3. Why are the elements carbon, hydrogen, oxygen, and nitrogen important to life on Earth?

4. How can there be so many different substances in the world if there are only a few elements that are common?

WARNING — This set contains chemicals that may be harmful if misused. Read cautions on individual containers carefully. Not to be used by children except under adult supervision.

Name ______________________

Period __________ Date ______________

HOW MUCH GAS? A

Materials for each group

- 1 Jar of sodium bicarbonate
- 1 Jar of citric acid
- 2 Spoons, 2-mL
- 1 Plastic cup, 250-mL
- 1 Stirring stick
- 2 Glass bottles
- 2 Rubber stoppers, #1, with hole
- 2 Syringes, 35-mL
- 1 Waste container
- 1 Tray or basin
- • Water
- • Protective eyewear

Procedure

a. Get a basin of *group materials* for your group.

b. Get a bottle-and-syringe system for each pair.

c. Put on protective eyewear.

d. Make a stock citric acid solution. Dissolve one level, 2-mL spoon of citric acid in 100 mL of water.

e. Put one level, 2-mL spoon of sodium bicarbonate into the bottle. Twist the stopper into the bottle.

f. Take up exactly 5 mL of citric acid solution in the syringe. Insert the tip of the syringe into the hole in the stopper.

g. Push the solution into the bottle. *Don't* remove the syringe. Observe and record.

h. Dump the used experiment and conduct two more trials. It is *not* necessary to wash out the bottle between trials.

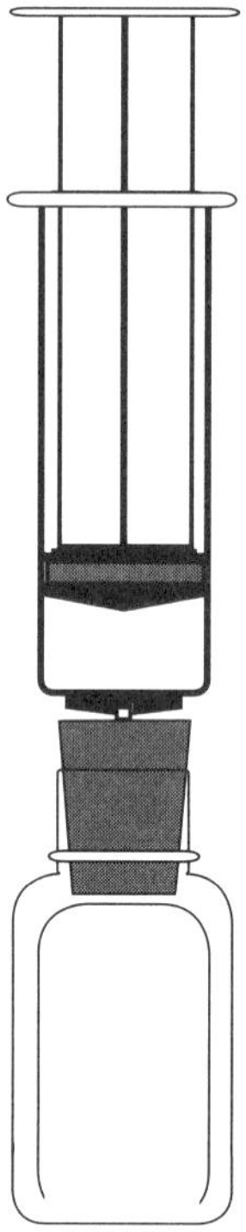

Volume of Gas Produced (mL)			
Trial 1	Trial 2	Trial 3	Average

Name ______________________________

Period ____________ Date ________________

HOW MUCH GAS? B

Analysis/Summary

1. What caused the syringe plunger to go up during the reaction between citric acid and sodium bicarbonate?

___

___

___

2. Why is a syringe more useful than a balloon to conduct this experiment?

___

___

___

3. What do you think might happen if you doubled the amount of citric acid solution?

___

___

Why do you think so?

___

___

4. What do you think might happen if you doubled the amount of sodium bicarbonate?

___

___

Why do you think so?

___

___

___

Name ______________________

Period __________ Date ______________

WHAT'S IN THE BUBBLES?

1. Make a list of the gases you know about or have heard about.

2. How would you define *gas*?

3. Everything is made of elements. What *elements* could be in the gas that forms when sodium bicarbonate ($NaHCO_3$) and citric acid ($C_6H_8O_7$) react?

4. What gas do you think is in the bubbles that form when $NaHCO_3$ and $C_6H_8O_7$ react?

Name ______________________________

Period ____________ Date ______________

DISCUSS AIR AS PARTICLES

1. What is the air in the syringe and the air in the bubble made of?

____

2. What happens to the air particles in the syringe when you push on the plunger?

____

3. What happens to the air particles in the bubble when you pull up on the plunger?

____

4. Are there more air particles in the bubble when it is compressed or when it is expanded?

____

5. When you push on the plunger, are the air particles closer together in the syringe or in the bubble?

____

6. What is between air particles?

____

7. What happens to air particles when a volume of air is compressed?

____

When a volume of air expands?

____

Name ______________________

Period ____________ Date ______________

AIR IN A SYRINGE A

A student had a syringe barrel. She drew a picture (A) of her idea of how air filled the room and the syringe.

She put the plunger into the barrel (B) and then clamped the syringe shut (C).

She pushed the plunger down (D) and pulled the plunger up (E).

Draw air particles in syringes B–E.

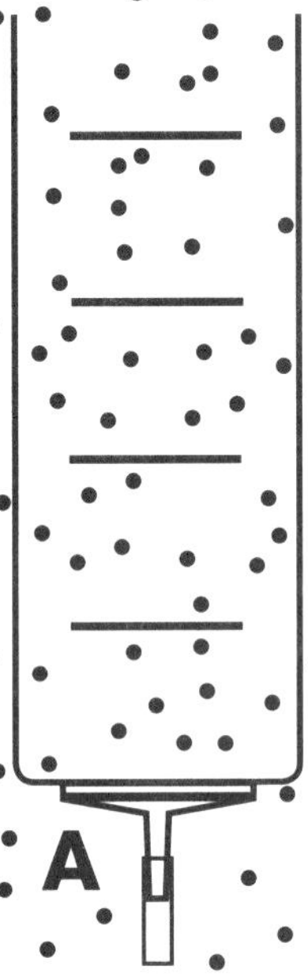

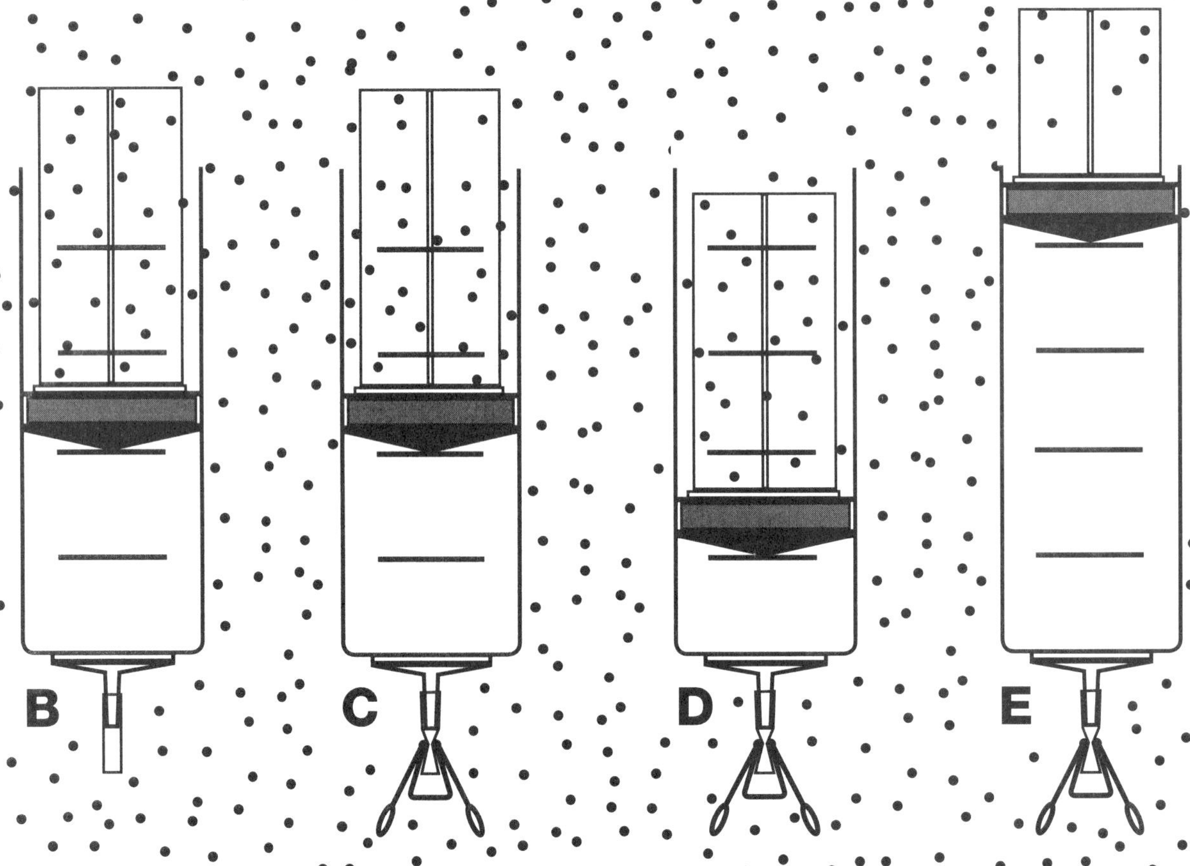

Name ______________________________

Period ____________ Date ________________

AIR IN A SYRINGE B

1. Why did you draw the particles in syringe B the way you did?

2. Why did you draw the particles in syringe C the way you did?

3. Why did you draw the particles in syringe D the way you did?

4. Why did you draw the particles in syringe E the way you did?

5. What happens to the air particles when air expands?

6. What happens to the air particles when air is compressed?

Name ______________________________

Period ____________ Date ______________

PARTICLES QUESTIONS

SRB page 15

1. What is a particle?

____

____

____

2. What is the difference between an element and a particle?

____

____

____

3. How many different kinds of particles are there in the world? Explain your answer.

____

____

____

Name ______________________________

Period ____________ Date ______________

THREE PHASES OF MATTER QUESTIONS

SRB page 22

1. What crumples a plastic bubble in a syringe when you apply force to the plunger?

2. How is the motion of particles in solid, liquid, and gas different?

3. Why does air feel hard when you push on the plunger of a closed syringe?

4. Explain why some foam cubes get smaller in a syringe and some stay the same size.

Name ______________________________

Period ___________ Date ______________

HEATING AND COOLING AIR A

Part 1. Question

What happens to a volume of air when it is heated? When it is cooled?

Part 2. Procedure

a. Work with materials to figure out a good demonstration to show fourth graders.

b. Draw and label your setup.

c. Write a description of what happens to air when it gets hot and when it gets cold. Make sure it can be understood by fourth graders.

Part 3. Draw and label your setup here.

Part 4. Explain what happens to air when it is heated and cooled.

Name ______________________________

Period ____________ Date ________________

HEATING AND COOLING AIR B

Part 5. Explain what happens at the particle level when air is heated and cooled.

- Imagine that you could see the air particles in the bottle.
- Explain what happens to the particles when the air is heated and cooled.
- Use drawings and labels if they will help.

____

____

____

____

____

____

____

____

Name ______________________

Period __________ Date __________

HEATING AND COOLING WATER A

Materials for each pair

- 1 Glass bottle
- 1 Rubber stopper with clear pipe
- 1 Syringe, 35-mL
- 1 Squeeze pipette
- 1 Card, 1″ × 3″
- • Tape
- • Blue water
- 1 Large cup (500 mL) with cold water
- 1 Large cup (500 mL) with hot water
- 1 Glass thermometer

Procedure

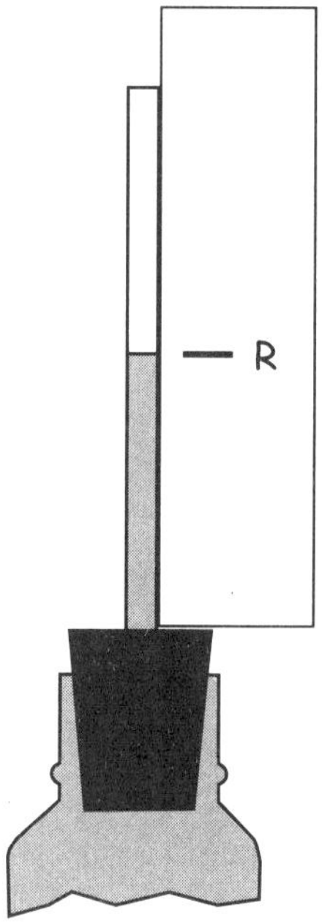

a. Push the clear plastic pipe a short distance into the rubber stopper.

b. Use a syringe to put 35 mL of blue water into the glass bottle.

c. Push the stopper into the bottle as far as it will go. Use the pipette to fine-tune the water level so it is halfway up the pipe.

d. Tape a 1″ × 3″ card to the clear tube. Label the water level "R."

e. Record the starting temperatures of the cold and hot water.

Cold water __________ Hot water __________

f. Place the bottle in cold water. After 3 minutes, label the water level "C."

g. Move the bottle to hot water. In 5 minutes, label the water level "H."

Think about the bottle system.

1. What happened when you placed your bottle system in cold water? Draw and explain.

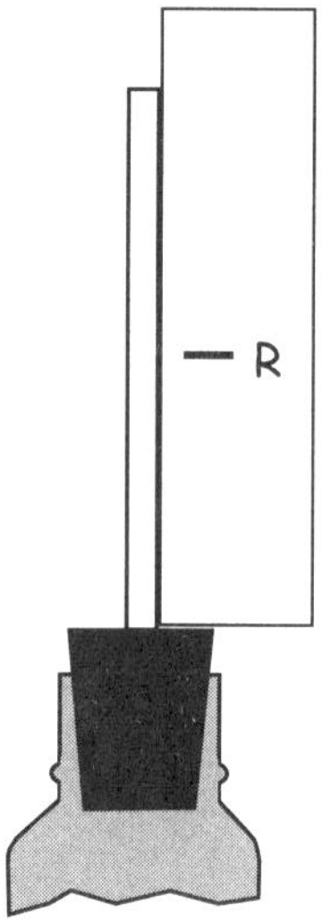

____

____

____

____

____

Name ______________________________

Period ____________ Date ______________

HEATING AND COOLING WATER B

2. What happened when you placed your bottle system in hot water? Draw and explain.

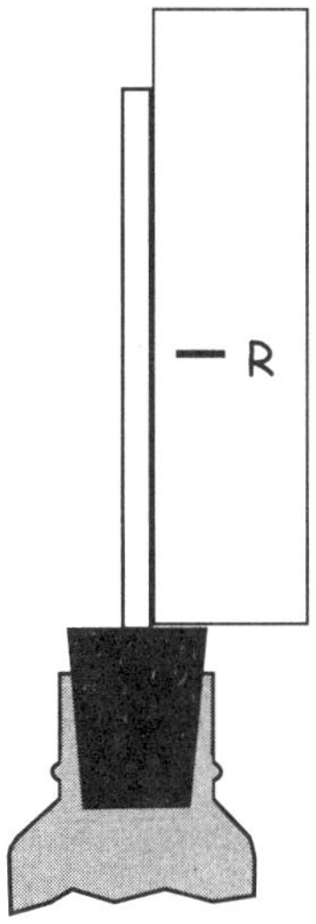

____

____

____

____

____

3. What caused the water to go up in the pipe when you put the bottle in hot water?

____

____

____

4. What caused the water to go down in the pipe when you put the bottle in cold water?

____

____

____

5. Describe what you think happened to the water particles in the bottle system when it was placed in hot water. Discuss kinetic energy and expansion.

____

____

____

____

____

Name ______________________________

Period ____________ Date ______________

PARTICLES IN MOTION QUESTIONS

SRB page 27

1. What is kinetic energy?

2. What are two ways to increase an object's kinetic energy?

3. Explain why a balloon inflates when a bottle-and-balloon system is placed in hot water.

4. What happens to a sample of matter when its particles lose kinetic energy?

5. How are particles in solids, liquids, and gases the same? How are they different?

Name ______________________________

Period __________ Date ______________

RESPONSE SHEET—KINETIC ENERGY

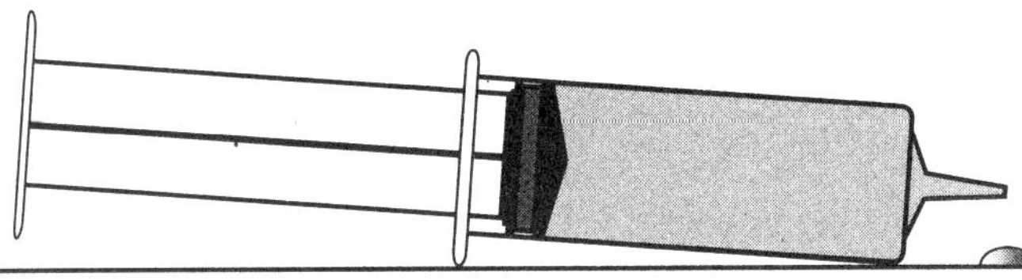

Bess filled a syringe with water and left it by the sink in the sunshine. Ten minutes later she saw a little puddle of water under the syringe tip. Bess said,

This syringe must be broken. It's leaking.

But it wasn't broken.

What do you think caused the little puddle of water to appear under the syringe tip?
NOTE: Use the words *particle* and *kinetic energy* in your explanation.

____

____

____

____

____

____

____

____

____

____

____

Name ______________________________

Period ____________ Date ______________

EXPANSION AND CONTRACTION QUESTIONS

SRB page 31

1. What are expansion joints, and why are they used?

____

____

____

____

2. What causes the cap to pop off a bottle of orange juice?

____

____

____

____

3. How does a thermometer work?

____

____

____

____

Name ______________________________

Period ____________ Date ____________

MIXING WATER

Question

If you mixed equal volumes of 50°C hot water and 10°C cold water, what do you think the temperature of the mixture would be?

Prediction

Predict the temperature of the mixture. ______________________________

Reasoning

Explain the thinking behind your prediction.

____

____

____

Procedure

Describe an experiment you can conduct to check your prediction.

____

____

____

____

____

Data

Conduct a water-mixing experiment.

We mixed ______ mL of hot water and ______ mL of cold water.

T_{hot} (°C)	T_{cold} (°C)	Prediction (°C)	T_{final} (°C)

Write the equation for calculating final temperature when equal volumes of water are mixed.

Name ______________________________

Period ____________ Date ________________

ENERGY ON THE MOVE QUESTIONS

SRB page 37

1. Explain how cold milk cools hot cocoa.

____

____

____

2. Why do you think an ice cube feels cold when you hold it in your hand?

____

____

____

3. What will happen to a balloon stretched over the mouth of an empty bottle when the bottle is placed in hot water? Explain all the energy transfers.

____

____

____

4. When does energy flow from a cold object to a hot object?

____

____

____

5. What does a thermometer measure, and how does it do it?

____

____

____

Name ______________________________

Period ____________ Date ______________

RESPONSE SHEET—ENERGY TRANSFER

Julie said,

When you put a bottle of juice in a cooler full of ice, the juice gets cold. That's because the cold transfers to the juice and slows down the kinetic energy of the juice particles.

Comment on Julie's ideas and give your explanation for why the juice gets cold.

Name ____________________

Period __________ Date __________

CALCULATING HEAT IN CALORIES A

Temperature is measured in degrees Celsius (°C). Heat is *not* measured in degrees Celsius. Heat is measured in **calories** (**cal**). The calorie is the unit of heat in the metric system.

One calorie is the amount of heat needed to raise the temperature of 1 g of water 1°C. For instance, it takes 1 cal of heat to raise the temperature of 1 g of water from 25°C to 26°C.

1. Calculate the number of calories needed to

	Calories (cal)
a. Raise the temperature of 1 g of water 1°C.	
b. Raise the temperature of 2 g of water 1°C.	
c. Raise the temperature of 2 g of water 2°C.	
d. Raise the temperature of 10 g of water 1°C.	
e. Raise the temperature of 1 g of water 70°C.	
f. Raise the temperature of 100 g of water 5°C.	
g. Raise the temperature of 450 g of water 3°C.	
h. Raise the temperature of 16 g of water 62°C.	

2. Billy mixed 40 g of 60°C water with 60 g of 25°C water. The final temperature was 39°C.

   a. Calculate the change of temperature (ΔT) for the hot water.

   $\Delta T = T_f - T_i$

   b. Calculate the amount of heat (calories) transferred *from* the hot water.

   calories = mass of hot water × *change of temperature of hot water*

   $cal = m \times \Delta T$

   c. Calculate the amount of heat transferred *to* the cold water.

   $cal = m \times \Delta T$

   d. Compare the amount of heat transferred *from* the hot water and the amount of heat transferred *to* the cold water.

Name ______________________________

Period ____________ Date ________________

CALCULATING HEAT IN CALORIES B

3. Rosella has a 10-liter fish tank. The water needs to be 28°C for her tropical fish. When she filled the tank, the temperature of the water was 12°C. How many calories of heat must transfer to the aquarium before it is ready for the fish?

4. Cindy made tea. She started with 300 g of water at 20°C. She transferred 18,000 cal to the water. What was the final temperature of the water?

5. Lee's 600-mL cup of cocoa got cold. It was only 25°C. He put it in the microwave. How many calories must transfer to the cocoa to bring it up to 70°C?

Name ______________________________

Period ____________ Date ________________

HEAT TRANSFER

Materials

2 Graduated cylinders
2 Pipettes
3 Foam cups
1 Large, clear cup (500-mL)
2 Thermometers
• Hot and cold water

Procedure

a. Decide on the mass of hot water and the mass of cold water you will use.

b. Measure the hot water into one foam cup and the cold water into a second foam cup.

c. Record the mass and starting temperatures in the table below.

d. Pour the cold water and the hot water into the third foam cup.

e. Put the third foam cup into the 500-mL cup.

f. Measure and record the final temperature.

Results

1	2	3	4	5	6
	Mass (g)	**Starting temp. (°C)**	**Final temp. (°C)**	**ΔT (°C)**	**Calories (cal)**
Hot water					
Cold water					

Conclusions

1. Calculate the calories transferred from the hot water. Show your math.

2. Calculate the calories transferred to the cold water. Show your math.

3. Compare the heat transfer from the hot water and the heat transfer to the cold water. What do you notice?

____

____

____

Name ____________________

Period __________ Date __________

HEAT PRACTICE A

1. What is the equation for calculating final temperature when equal masses of water are mixed?

2. What is the equation for calculating how much heat energy (calories) transferred to or from a mass of water?

3. Mix 30 mL of water at 15°C and 30 mL of water at 55°C.

Answer these questions. Show your work.

a. What is the final volume of the water?

b. What is the final temperature of the water?

c. How many degrees did the cold water increase?

d. How many degrees did the hot water decrease?

e. How much heat energy transferred to the cold water?

f. How much heat energy transferred from the hot water?

g. What happened to the kinetic energy of the hot-water and cold-water particles?

Name ______________________________

Period ____________ Date ______________

HEAT PRACTICE B

4. Mix 25 mL of water at 0°C and 25 mL of water at 50°C.

Answer these questions. Show your work.

a. What is the final temperature of the water?

b. What is the ΔT for the cold water?

c. How much heat energy transferred to the cold water?

d. What is the ΔT for the hot water?

e. How much heat energy transferred from the hot water?

5. Energy "flow" is the transfer of energy from one place to another. Which direction does energy flow?

____

____

6. How does energy transfer happen?

____

____

7. What is equilibrium?

____

____

Name ______________________

Period __________ Date __________

ICE WATER AND HOT WATER A

Question

Can you predict the final temperature of a mixture of ice water and hot water?

Materials

2 Foam cups
1 Glass thermometer
1 Stirring stick
1 Ice cube, 30–40 g
• Hot water
• Cold water
• Balance

Procedure

a. Weigh an ice cube in a foam cup. Record its mass in the table below.
b. Add ice water (0°C) to bring the total mass of ice and water up to 60 g.
c. Measure the temperature of the ice water. Record.
d. Put 60 g of hot water in a second foam cup.
e. Measure the temperature of the hot water. Record.
f. Predict the equilibrium temperature of the mixture. → [ ]
g. Pour the hot water into the ice water and stir gently until the ice is all melted.
h. *When the ice is melted,* record the final temperature.

Results

Material	Mass (g)	T_i (°C)	T_f (°C)	ΔT (°C)	Calories (cal)
Ice					
Ice/water mix					
Hot water					

Calculate $cal = m \times \Delta T$

a. Calculate the calories transferred from the hot water. Show your math.

b. Calculate the calories transferred to the ice and ice water. Show your math.

Name ______________________________

Period ____________ Date ______________

ICE WATER AND HOT WATER B

Conclusions

1. Compare the observed equilibrium temperature with your prediction. Were they close?

2. How many calories transferred *from* the hot water?

3. How many calories transferred *to* the cold water and ice?

4. Is the energy (calories) transferred *from* the hot water balanced by the energy (calories) transferred *to* the cold water? Show your math.

5. What do you think caused the low final temperature of the mixture?

____

____

____

____

____

6. What do the results of the investigation suggest about energy (calories) and ice?

____

____

____

____

____

Name ______________________

Period __________ Date __________

HEAT OF FUSION A

Some scientists landed on a planet that has an ocean of liquid tarpoo with solid chunks of tarpoo floating in it. The tarpoo was found to melt at 20°C. The scientists mixed solid and liquid tarpoo and took the temperature when the solid had all melted. Use their data to determine the heat of fusion of tarpoo.

Material	Mass (g)	Starting temp. (*T*)	Ending temp. (*T*)	Change of temp. (ΔT)	Energy transfer (cal)
Solid tarpoo	100	20°C	40°C	20	2,000
Liquid tarpoo	100	100°C	40°C	60	6,000

NOTE: 1 calorie (cal) = the amount of heat needed to change the temperature of 1 g of liquid tarpoo 1°C.

Name ______________________

Period __________ Date __________

HEAT OF FUSION B

Some scientists landed on a planet that has lakes of liquid grisk with pieces of solid grisk around the edge. The grisk was found to melt at 25°C. The scientists mixed solid and liquid grisk and took the temperature when the solid had all melted. Use their data to determine the heat of fusion of grisk.

Material	Mass (g)	Starting temp. (*T*)	Ending temp. (*T*)	Change of temp. (ΔT)	Energy transfer (cal)
Solid grisk	50	25°C	45°C	20	1,000
Liquid grisk	50	125°C	45°C	80	4,000

NOTE: 1 calorie (cal) = the amount of heat needed to change the temperature of 1 g of liquid grisk 1°C.

Name ______________________________

Period ____________ Date ______________

HEAT OF FUSION C

Some scientists landed on a planet that has large lakes of liquid neotrene with chunks of solid neotrene around the edges. The neotrene was found to melt at 15°C. The scientists mixed solid and liquid neotrene and took the temperature when the solid had all melted. Use their data to determine the heat of fusion of neotrene.

Material	Mass (g)	Starting temp. (*T*)	Ending temp. (*T*)	Change of temp. (ΔT)	Energy transfer (cal)
Solid neotrene	100	15°C	40°C	25	2,500
Liquid neotrene	100	95°C	40°C	55	5,500

NOTE: 1 calorie (cal) = the amount of heat needed to change the temperature of 1 g of liquid neotrene 1°C.

Name ______________________________

Period ____________ Date ______________

HEAT OF FUSION D

Some scientists landed on a planet that has pools of liquid simgob with crystals of solid simgob around the edge. The simgob was found to melt at 15°C. The scientists mixed solid and liquid simgob and took the temperature when the solid had all melted. Use their data to determine the heat of fusion of simgob.

Material	Mass (g)	Starting temp. (*T*)	Ending temp. (*T*)	Change of temp. (ΔT)	Energy transfer (cal)
Solid simgob	200	15°C	35°C	20	4,000
Liquid simgob	200	100°C	35°C	65	13,000

NOTE: 1 calorie (cal) = the amount of heat needed to change the temperature of 1 g of liquid simgob 1°C.

Name ______________________________

Period ____________ Date ______________

HEAT OF FUSION QUESTIONS

SRB page 41

1. What is heat of fusion?

2. What happens at the particle level when you put ice cubes in a glass of room-temperature lemonade?

3. Explain how an ice chest cools a can of soft drink.

DISSOLVE OR MELT? A

Materials

2 Plastic cups
2 Aluminum foil squares
2 Paper cups
4 Candies, all one color
- Hot water
- Cold water

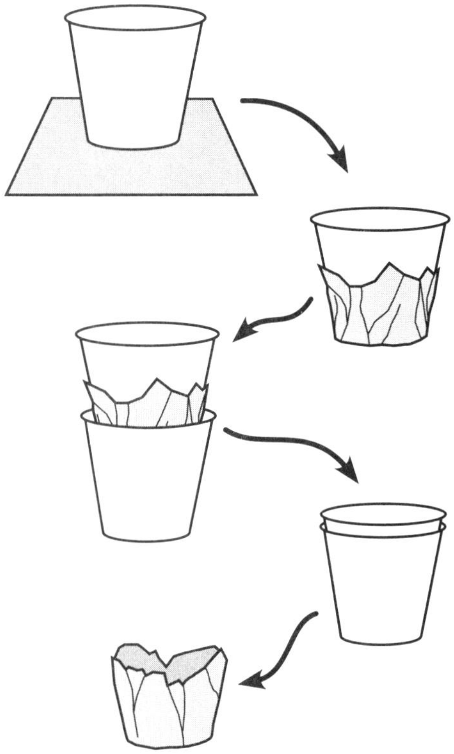

Prepare foil cups

a. Place a plastic cup in the center of a foil square. Bring the foil up around the edges of the cup.

b. Place the foil-wrapped cup inside a second cup. Push gently but firmly all the way down.

c. Remove the foil cup. The foil cup will float on the water in a plastic cup.

d. Repeat the procedure to make a second aluminum foil cup.

Procedure

a. Put about 150 mL of hot water in one plastic cup; put about 150 mL of cold water in the other plastic cup.
b. Put an aluminum foil cup in each cup of water.
c. Get four candies, all one color. Put one candy in each aluminum foil cup and one in the bottom of each of the cups of water.
d. Don't stir, poke, or shake the candies or the cups. Observe to see if anything melts and if anything dissolves.

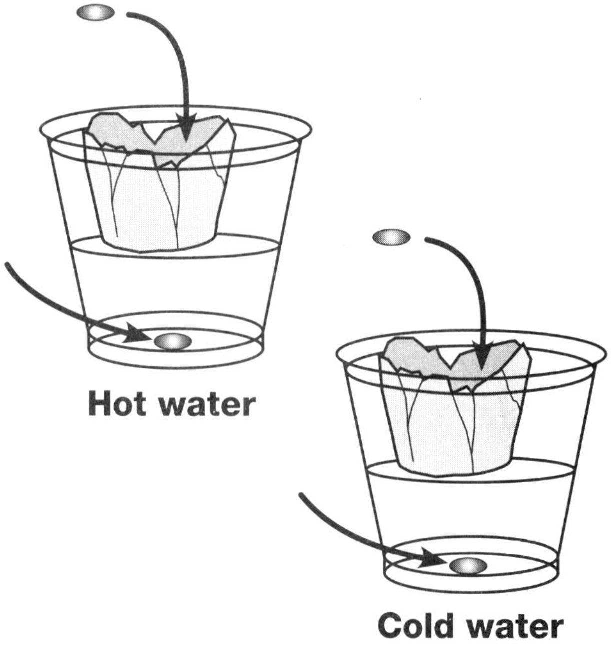

Name ______________________________

Period ____________ Date ________________

DISSOLVE OR MELT? B

Results

Record your observations in the table.

Material	Hot water	Cold water	Hot air	Cold air
Candy coating				
Chocolate				

Conclusions

1. a. What melted? ______________________________

   b. Under what conditions? ______________________________

   c. What happened at the particle level when it melted?

   ____

   ____

2. a. What dissolved? ______________________________

   b. Under what conditions? ______________________________

   c. What happened at the particle level when it dissolved?

   ____

   ____

   ____

Name ______________________________

Period ____________ Date ________________

MELT THREE MATERIALS

Materials

- Containers, 1/2-liter
- Paper cups
- Thermometers
- Margarine cubes
- Wax chunks
- Sugar
- Hot water

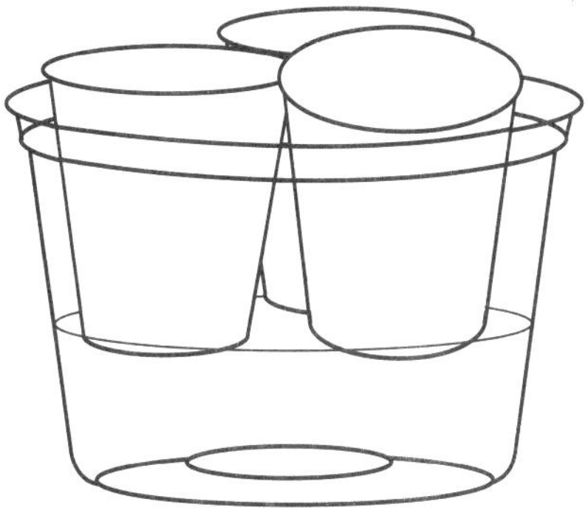

Prediction

Will margarine, wax, and sugar melt in hot water? Record your predictions in the table below. Then write your procedure and conduct the test.

Procedure

____

____

____

____

____

____

____

Results

Material	Prediction: Will it melt?	Water temperature (°C)	Observations
Margarine			
Wax			
Sugar			

Name ______________________________

Period ____________ Date ________________

WAX AND SUGAR QUESTIONS

1. What does the word *melt* mean?

2. Did the wax melt?

3. What was your evidence?

4. Did the sugar melt?

5. What was your evidence?

6. Did the melted wax and sugar stay liquid?

7. Did the melted wax and sugar freeze? What is your evidence?

8. Do all solids melt? How would you find out?

9. Do all solids melt at the same temperature?

10. Do all liquids freeze at the same temperature?

11. How could you find out if all liquids freeze?

12. When wax melts, how do the wax particles change?

13. Why do materials melt when they get hot?

14. What happens at the particle level when a material freezes?

15. Look at the puddle of wax around the wick of your candle. Explain why it is solid now.

Name ______________________________

Period __________ Date ______________

ROCK SOLID QUESTIONS

SRB page 48

1. What causes a substance to change from one phase to another?

2. What are the three important things to know about freezing and melting?

3. Why does liquid water form on the bottom of a cup of ice placed over warm water?

4. What happens to water particles as a cup of ice melts and then evaporates?

Name ______________________________

Period ____________ Date ______________

RESPONSE SHEET—PHASE CHANGE

Randy watched his mom put a piece of wax in a pan. She put the pan on the stove. A minute later, Randy looked in the pan and said,

Look, the wax is turning into water.

What would you tell Randy to help him understand what happened in the pan?

____

____

____

____

____

____

____

____

____

____

____

____

WARNING — This set contains chemicals that may be harmful if misused. Read cautions on individual containers carefully. Not to be used by children except under adult supervision.

Name ______________________________

Period ____________ Date ______________

FREEZE WATER A

Materials

- 1 Glass thermometer
- 1 Metal-backed thermometer
- 1 Plastic cup
- 1 Vial
- 1 Stirring stick
- • Sodium chloride, 3 spoons
- • Ice, crushed
- • Protective eyewear
- • Water

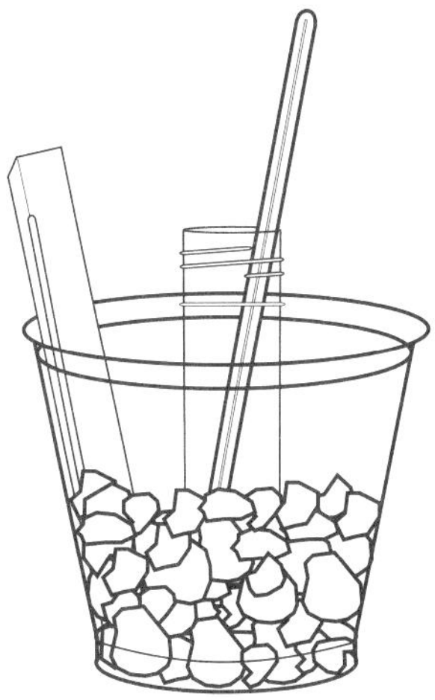

Procedure

a. Fill a plastic cup halfway with crushed ice.

b. Put on protective eyewear. Add three 5-mL spoons of sodium chloride to the ice. Stir in thoroughly.

c. Put about 10 mL of water in a vial.

d. Carefully work the vial of water into the crushed ice. Make sure the surface of the water is below the level of the ice.

e. Monitor the temperature of the water in the vial with a glass thermometer. Monitor the temperature of the ice/salt environment with a metal-backed thermometer.

f. Record your observations. Include time, temperatures, and changes to the system.

Results

Time	Water temperature (°C)	Ice bath temperature (°C)	Observations

Name

Period   Date

FREEZE WATER B

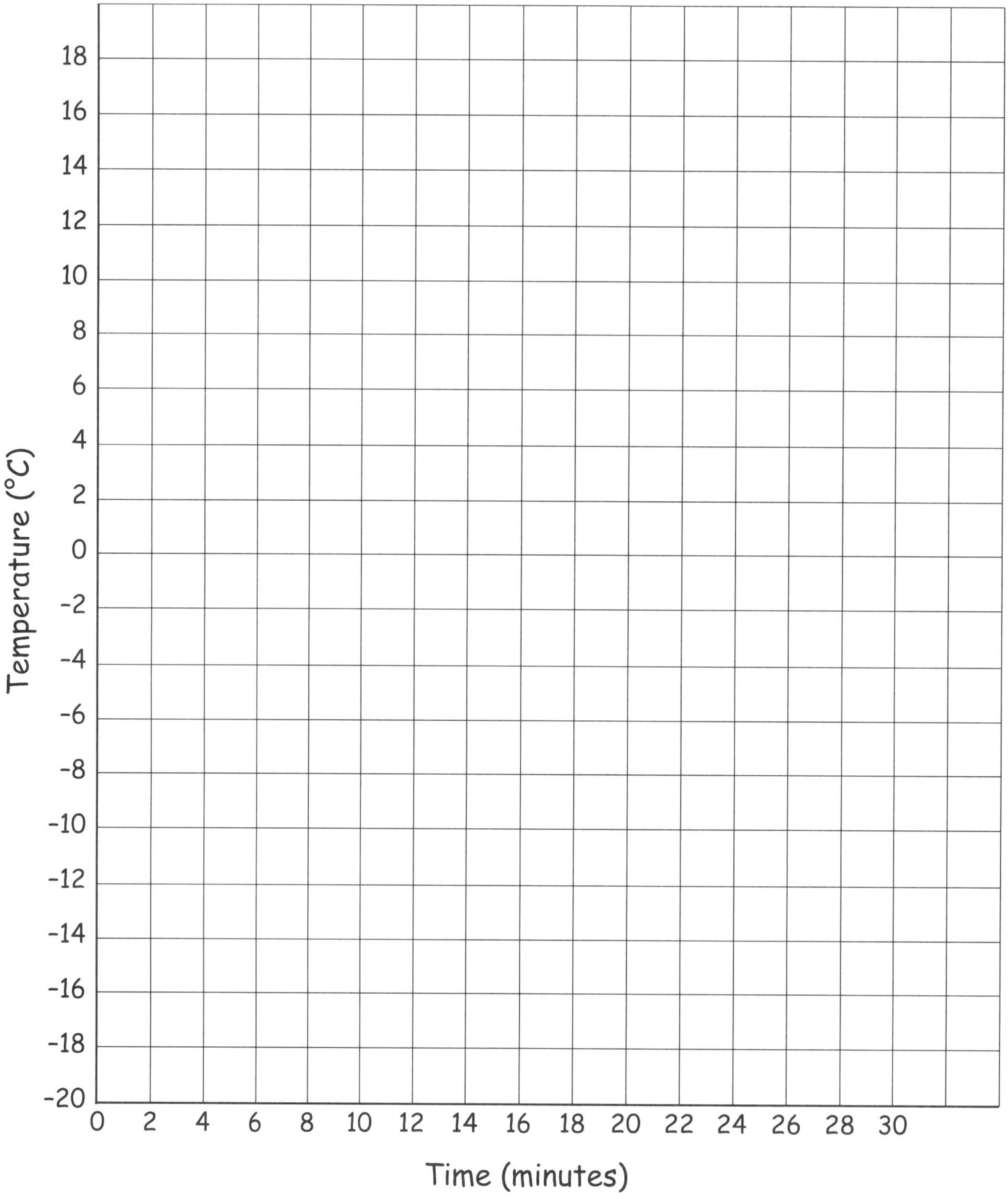

Name ______________________

Period __________ Date ______________

FREEZE WATER C

Conclusions

1. Describe what happened to the ice / salt mixture as the investigation progressed.

2. Describe what happened to the vial of water as the investigation progressed.

3. What happened to the temperature of the water in the vial as the water was freezing?

4. Why do you think the vial of water in plain ice didn't freeze, but the vial of water in salted ice did freeze?

5. People put salt on ice when they make ice cream. Why do they do that?

Name ______________________________

Period ____________ Date ______________

WATER-AND-ICE SYSTEM OBSERVATIONS

Write a description of the changes you observed when a cup of ice was placed over warm water and then when salt was added to the ice.

Include particles, energy transfer, and phase change in your description, and label the illustration.

WARNING — This set contains chemicals that may be harmful if misused. Read cautions on individual containers carefully. Not to be used by children except under adult supervision.

Name ______________________________

Period ____________ Date ______________

MIXTURES A

Materials

- 2 Plastic cups, 250-mL
- 2 Self-stick notes
- 2 Stirring sticks
- 2 Hand lenses
- • Protective eyewear
- 1 Syringe, 35-mL
- 1 Container of water
- • Calcium carbonate ($CaCO_3$)
- • Sodium chloride (NaCl)

Procedure

a. Label two cups using self-stick notes: "Calcium carbonate ($CaCO_3$)" and "Sodium chloride (NaCl)."

b. Put on protective eyewear.

c. Measure one level, 2-mL spoon of calcium carbonate into one plastic cup.

d. Measure one level, 2-mL spoon of sodium chloride into a second plastic cup.

e. Observe the two solid materials with a hand lens. Record your observations.

f. Use a syringe to add 30 mL of water to each cup. Stir, observe, and record.

Observations

Substance	Before mixing with water	After mixing with water

WARNING — This set contains chemicals that may be harmful if misused. Read cautions on individual containers carefully. Not to be used by children except under adult supervision.

Name ______________________

Period __________ Date ______________

MIXTURES B

Materials for separating mixtures

- 2 Plastic cups, 250-mL
- 2 Self-stick notes
- 1 Funnel stand
- 2 Filter papers, small
- 2 Hand lenses
- 1 Well tray
- 3 Pipettes
- • Protective eyewear

Filtering procedure

a. Label two cups: "Calcium carbonate ($CaCO_3$)" and "Sodium chloride (NaCl)."

b. Set up the funnel stand and place a filter paper in the funnel.

c. Place the empty cup labeled sodium chloride under the filter funnel. Pour the sodium chloride mixture into the filter.

d. Repeat the process with the calcium carbonate mixture.

Filtering results

____

____

____

Follow-up procedures

____

____

____

____

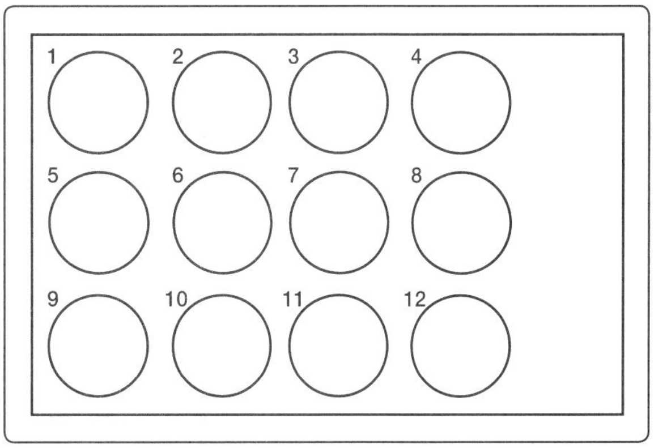

Follow-up results

____

____

____

Name ______________________________

Period ____________ Date ________________

HOW THINGS DISSOLVE QUESTIONS

SRB page 53

1. Copper chloride ($CuCl_2$) dissolves in water. Describe what happens at the particle level when copper chloride is put into water.

____

____

____

2. What are some of the solutions found in living organisms?

____

____

____

3. Is milk a mixture, a solution, or both? Why do you think so?

____

____

____

4. How could a solution of copper chloride and water be separated into its starting substances?

____

____

____

WARNING — This set contains chemicals that may be harmful if misused. Read cautions on individual containers carefully. Not to be used by children except under adult supervision.

Name ______________________________

Period ____________ Date ________________

HOW MUCH WILL DISSOLVE? A

Dissolving questions

1. Is there a limit to the amount of sodium chloride that will dissolve in 30 g of water?
2. If there is a limit, how much sodium chloride will dissolve in 30 g of water?
3. Will the same amount of magnesium sulfate dissolve in 30 g of water?

Materials available

- Plastic bottles with caps
- Self-stick notes
- Jars of NaCl
- Jars of $MgSO_4$
- Spoons, 5-mL
- Syringes, 35-mL
- Plastic cups, 250-mL
- Funnel stand
- Filter papers, large
- Balances
- Water
- Protective eyewear

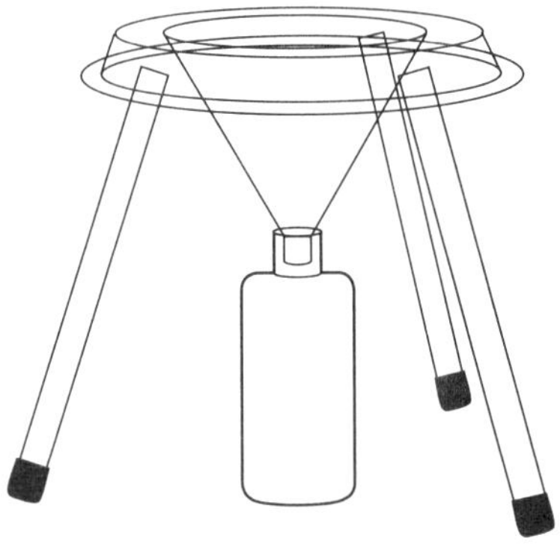

Experimental procedure

Design an experiment to answer the questions. Get approval from your teacher before starting your experiment.

WARNING — This set contains chemicals that may be harmful if misused. Read cautions on individual containers carefully. Not to be used by children except under adult supervision.

Name ______________________________

Period ____________ Date ______________

HOW MUCH WILL DISSOLVE? B

Results

Solvent	Mass of solvent (g)	Solute	Mass of solution (g)	Mass of solute (g)

Conclusions

What do the results of the two saturation experiments tell you?

Deeper thoughts

What do you think happened in your solution bottle at the particle level?

Name ________________________________

Period ____________ Date ________________

RESPONSE SHEET—SOLUTIONS

Jill's mom made a cup of hot tea and dropped in a sugar cube. In 30 seconds, it was gone.

Jill's mom said,

Look, this tea was so hot the sugar melted.

Jill said,

I don't think so. I think it dissolved.

Who do you think is right? Describe what you think happened at the particle level when the sugar cube was placed in the hot tea.

WARNING — This set contains chemicals that may be harmful if misused. Read cautions on individual containers carefully. Not to be used by children except under adult supervision.

Name ______________________

Period ____________ Date ______________

MAGNESIUM SULFATE SOLUTIONS

Recipe	Magnesium sulfate (g)	Water (g)		
1	5	25		
2	8	20		
3	12	20		
4	8	40		
5	10	25		
6	15	25		
7	10	50		
8	16	40		

1. What is the order of concentration of the magnesium sulfate solutions? Why do you think so?

____

____

____

____

____

____

2. Why do you think 20 mL of magnesium sulfate solution has a greater mass than 20 mL of plain water?

____

____

____

____

____

____

Name ______________________________

Period ____________ Date ____________

CONCENTRATION QUESTIONS

SRB page 62

1. What is the difference between a concentrated solution and a dilute solution?

____

____

____

2. Why does juice taste "weak" after the ice in it melts?

____

____

____

3. How can 50 mL of one salt solution have more mass than 50 mL of a second salt solution?

____

____

____

4. What is the maximum concentration of mercury allowed in drinking water in the United States?

____

____

____

Name ____________________

Period __________ Date __________

REPRESENTING SUBSTANCES

Substance name	Chemical formula	Representation	Number of elements	Number of atoms
Carbon	C			
Water	H_2O			
Carbon dioxide	CO_2			
Sodium chloride	NaCl			
Oxygen	O_2			
Sodium carbonate	Na_2CO_3			

Name ______________________________

Period ____________ Date ______________

ANALYZING SUBSTANCES

Substance name	Chemical formula	Representation	Number of elements	Number of atoms

		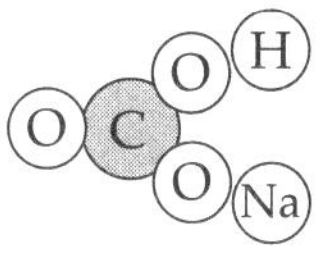		
		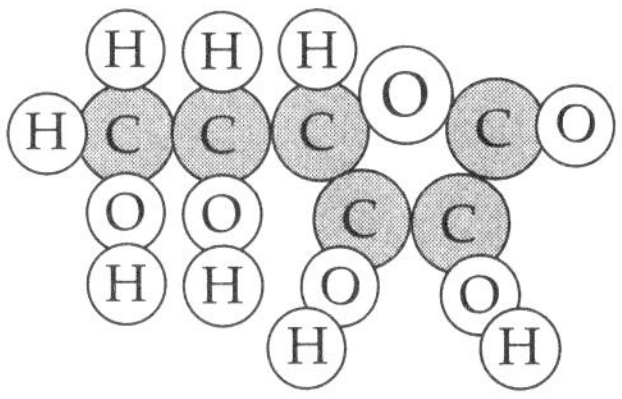		
		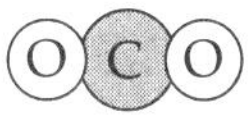		
		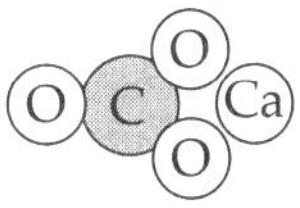		
		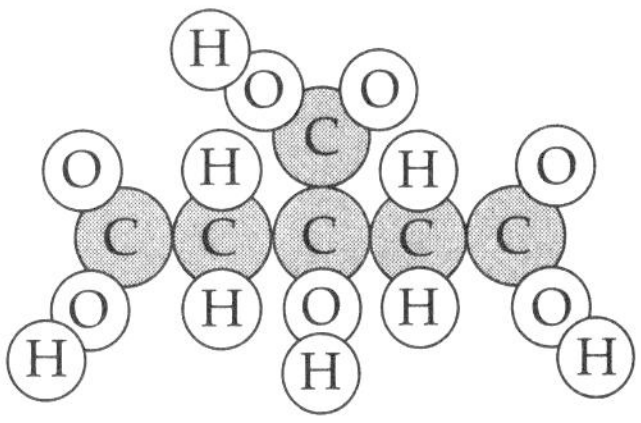		
		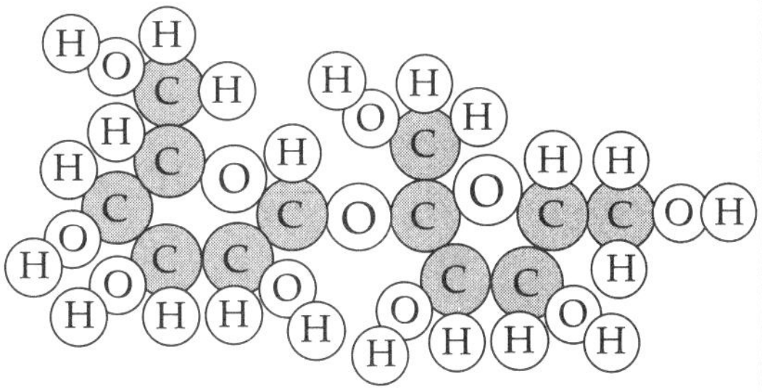		

WARNING — This set contains chemicals that may be harmful if misused. Read cautions on individual containers carefully. Not to be used by children except under adult supervision.

Name ____________________

Period __________ Date __________

LIMEWATER INVESTIGATION A

Materials

- Limewater (calcium hydroxide solution)
- 1 Plastic bottle
- 1 Rubber stopper, 2-hole
- 2 Clear plastic pipes
- 1 Short piece of tubing
- 1 Long piece of tubing
- Straw mouthpieces
- Protective eyewear

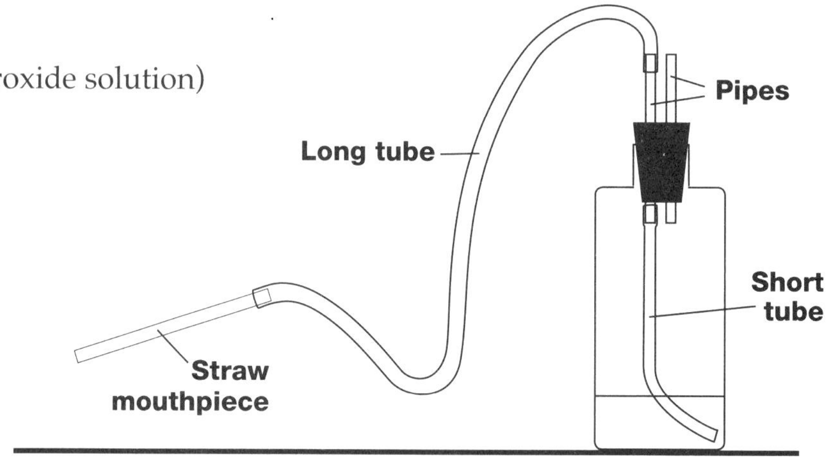

Procedure

a. Push the two clear plastic pipes through the holes in the rubber stopper.

b. Attach a long piece of tubing and a short piece of tubing to one pipe, as illustrated.

c. Put on protective eyewear. Measure 30 mL of limewater into the bottle. Insert the rubber stopper in the bottle.

d. Take turns using your straw mouthpieces to *gently* bubble one breath of air into the bottle through the long tube. Everyone should have at least two turns.

Results

Describe the changes you observed in the bottle.

___

___

___

Conclusion

Starting substances change into new substances during chemical reactions. Do you think a reaction occurred in the bottle? Why or why not?

___

___

___

___

Name ______________________

Period __________ Date ______________

LIMEWATER INVESTIGATION B

Thinking about limewater

1. Limewater is calcium hydroxide dissolved in water. The chemical formula for calcium hydroxide is $Ca(OH)_2$. Use circles labeled with atomic symbols to draw what you think a representation of one particle of calcium hydroxide might look like.

2. a. Use atom tiles to make representations of the particles that you think reacted.

   b. Rearrange the atoms to figure out what the white precipitate is.

   c. Draw representations of the reactants and the products using labeled circles.

   (**HINT**: The white powder does not dissolve in water.)

3. Write the limewater reaction using chemical formulas. Write the names of the reactants and products under the formulas.

   $Ca(OH)_2$ +

   Calcium hydroxide

4. Did new substances form? ________ If yes, what are they? ______________________
5. Did new atoms form? ____________ If yes, what are they? ______________________
6. Did new elements form? __________ If yes, what are they? ______________________

Name ____________________

Period __________ Date __________

HOW DO ATOMS REARRANGE? QUESTIONS

SRB page 68

1. What is destroyed and what is created during chemical reactions?

2. What are reactants and products? Write a reaction equation and label the reactants and products.

3. Write the equation for the reaction between hydrogen and oxygen. Use chemical formulas for the substances.

4. Methane (CH_4) is the main gas in natural gas. The products that form when methane burns are carbon dioxide and water. Write a balanced equation showing the combustion reaction when methane and oxygen react.

WARNING — This set contains chemicals that may be harmful if misused. Read cautions on individual containers carefully. Not to be used by children except under adult supervision.

Name ______________________________

Period ____________ Date ______________

ACID/SODA REACTION PRODUCTS

Materials

1 Bottle, 120-mL
1 Syringe, 35-mL
1 Midispoon
1 Stopper-and-tubes setup

1 Vial
1 Vial holder
1 Well tray
1 Pipette

- Hydrochloric acid (HCl)
- Baking soda ($NaHCO_3$)
- Limewater ($Ca(OH)_2$)
- Protective eyewear

Procedure

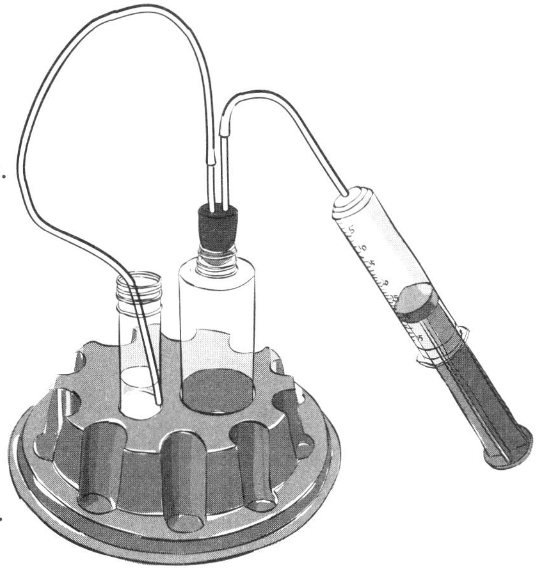

a. Put on protective eyewear.

b. Place 3 *level* midispoons of baking soda in the bottle.

c. Place about 10 mL of limewater in the vial.

d. Insert the bottle and the vial into the cavities in the center of the vial holder.

e. Take up *5 mL* of hydrochloric acid in the syringe.

f. Draw *30 mL* of air into the syringe.

g. Slowly put the acid and air into the bottle. Observe.

Results

1. What happened in the *bottle*? Use chemical equations to explain.

2. What happened in the *vial*? Use chemical equations to explain.

3. Were you able to confirm all the products that formed during the reaction between baking soda and hydrochloric acid? If not, what else will you do?

____

____

____

Name ______________________________

Period ____________ Date ______________

RESPONSE SHEET—REACTION

Grandmother ate too many chili peppers for supper. She moaned,

I need an antacid tablet.

Beth found the package of antacid tablets and read the label. The active ingredient was calcium carbonate. Beth said,

This will give you some relief.

1. Explain why Beth thought the antacid tablet would help Grandmother.

____

____

____

____

____

____

____

____

2. Use chemical formulas to write the equation for the reaction.

NOTE: Here are the formulas for some of the substances you have used.

CO_2	H_2O	Na_2CO_3
$CaCO_3$	HCl	$NaCl$
$CaCl_2$	$MgSO_4$	
$Ca(OH)_2$	$NaHCO_3$	

Name ______________________

Period __________ Date __________

LAVOISIER QUESTIONS

SRB page 72

1. Why did mercuric oxide in Lavoisier's reaction chamber weigh more than the mercury metal?

2. Why was there less air in Lavoisier's reaction chamber after he heated the mercury for 12 days?

3. What are some of the reasons Lavoisier is considered to be the father of modern chemistry?

WARNING — This set contains chemicals that may be harmful if misused. Read cautions on individual containers carefully. Not to be used by children except under adult supervision.

Name ______________________________

Period ____________ Date ______________

HEARTBURN CHEMISTRY

Question

How much stomach acid can one antacid tablet neutralize?

Materials

- 1 Plastic cup, 250-mL
- 1 Antacid tablet
- 1 Syringe, 35-mL, for measuring acid
- • Hydrochloric acid (HCl)
- • Protective eyewear

Procedure

____

____

____

____

____

Conclusions

1. How many milliliters of acid does one antacid tablet neutralize? ______________

2. The hydrochloric acid used in class is about 10 times more concentrated than real stomach acid. How many milliliters of real stomach acid will one antacid tablet neutralize? Show your math.

3. Write the chemical equation for the reaction between hydrochloric acid and the antacid.

Name ______________________________

Period ____________ Date ______________

CITRIC ACID/BAKING SODA REACTION

5 mL of citric acid solution A + **3 midispoons** of baking soda → about 30 mL of gas

5 mL of citric acid solution A + **6 midispoons** of baking soda → about 30 mL of gas

1. Explain why.

____

____

____

5 mL of citric acid solution **A** + 3 midispoons of baking soda → about 30 mL of gas

5 mL of citric acid solution **B** + 3 midispoons of baking soda → about 15 mL of gas

2. Explain why.

____

____

____

3. Discuss what you think might happen if you mixed 5 mL of citric acid solution A with **1 midispoon** of baking soda.

____

____

____

4. Explain what you could do to determine which reactant was in excess at the end of a reaction between citric acid and baking soda.

____

____

____

____

____

____

Name ______________________________

Period ____________ Date ________________

RUST

Observations

1. Describe the steel wool in the cylinder.

____

____

2. Describe the air in the cylinder.

____

____

Results

1. Write a balanced equation for the reaction between iron (Fe) and oxygen (O_2) to form rust (Fe_2O_3).

2. Explain why the water level inside the cylinder changed.

____

____

____

3. Has the steel wool stopped rusting or is it continuing to rust? Explain.

____

____

____

4. Explain which reactant is the limiting factor in the steel wool experiment.

____

____

____

NOTES

NOTES

NOTES

NOTES